Copyright©

All rights reserved. This book or parts thereof may not be reproduced in any form, stored in any retrieval system, or transmitted in any form by any means—electronic, mechanical, photocopy, recording, or otherwise—without prior written permission of the publisher except in the case of brief quotations embodied in critical articles or reviews.

TABLE OF CONTENTS

INTRODUCTION 4

HOW TO MARKET YOUR BUSINESS ON TWITTER 5

- How to Start a Twitter Account for Your Business 6
- How to create a Twitter marketing strategy for your brand 16
- How to use Twitter for marketing ... 17
- Twitter global times to post 2021 ... 24
- Pros of Using Twitter For Business Marketing 38
- Cons & Disadvantages For Using Twitter For Business 39
- Effective Twitter Marketing tips for your Business success 41

USING THE BEST HASHTAGS FOR YOUR SOCIAL STRATEGY. 66

What is a hashtag? 67

Hashtag best practices 68

How to find the best hashtags 69

How many hashtags should you use? 71

How to track your hashtags........... 74

Which hashtags to track 77

Twitter analytics tools to amplify your strategy...................................... 82

What comes next?....................... 91

CONCLUSION 93

INTRODUCTION

Twitter is a highly-efficient social media marketing tool for all types of businesses, and it is the perfect platform to promote your business. If you use this platform effectively, you can increase brand recognition, increase sales, and expand the reach. It is a great platform for customer service as well.

HOW TO MARKET YOUR BUSINESS ON TWITTER

Before diving into the topic, let's have a look at some important statistics that actually give a clear view of why you should use Twitter for your business.

• Twitter has over 330 million active monthly users.

• 75% of B2B businesses are marketing on Twitter and 65% of B2C businesses are marketing on Twitter.

• 79% of Twitter users more likely to recommends brands that they follow.

• 67% of Twitter users more likely to buy from brands that they follow.

• 85% of users feel more connected with businesses after following them.

- 40% of Twitter users say they have made a purchase as a result of Tweet from an influencer.

That's why it said that Twitter is a highly- efficient marketing tool to grow your business.

Do you know?

The average life span of a tweet is extremely short and Moz found that the average life span of one tweet is only 18 minutes and 500 million tweets are sent every day. Over 8500 new tweets are sent every single second.

This is extremely crowded, right?

Then, how will you use Twitter for your business effectively?

How to Start a Twitter Account for Your Business

One of the most powerful tools for businesses is still underutilized in today's market. And that tool is Twitter. The significant push toward both social media and branding has made it nearly impossible to avoid this social space.

Whether you're educating customers about your product or service, reaching a new audience or promoting your brand, Twitter is one of the most useful places to be to achieve your marketing goals.

When brands steer clear of Twitter, they usually use the excuse that they don't know how to use it. Businesses also feel the network can be overwhelming for a single person. While some marketers might not have room on their plate to manage multiple social media accounts, there's no excuse for being inactive on Twitter as a business.

To ensure you get off on the right foot, here are six simple and easy-to-follow steps on how to start a Twitter account for your business:

1. Make a Twitter Account

While this is the most obvious step, simply making a Twitter account is a crucial part to being active on social media. At Twitter's site, you just enter your name, work email and create a multi-user friendly password for the account. Once you've signed up, you begin filling out some information for the rest of your account.

Don't be afraid to enter your name and work email on certain questions. These features are changeable and easy to format for your business. Once you're ready to pick a Twitter Handle, remember the shorter the better. You want to be searchable for your customers and others in the industry.

You want to get into the practice of making everything short and concise on Twitter because you only have 140 characters to use. Additionally, you want to think about when customers or other users reach out to. When they @mention you or reply to one of your posts, your username will take up some of the characters for the response. It's always best to keep usernames limited for this reason.

Here are a few things to remember when signing up:

• You can change your username later.

• Your Twitter password should contain letters, numbers and symbols.

• Confirm your account before trying to personalize.

• Log out of your personal Twitter account before creating a new one.

2. Connect With Others

Like any social media network, it's all about connecting with other users to share content. Luckily, Twitter is great spot for new users to build connections and the network even gives you a head start to find influencers.

Twitter allows you to connect with others as soon as you fill out your profile. You'll receive recommendations on who to follow, whether it's celebrities, athletes, comedians, business leaders or musicians. However, you can browse for people that would match your business's interests.

Once you begin to add people, Twitter provides helpful suggested users to follow based on your previous selections. Additionally, you can see related users when you're on someone's feed you follow. The process is extremely easy and unlike Facebook, Twitter is a great spot to add people you don't quite know. The social network is more about building connections than intimate or personal networks.

3. Add Profile Details

So you've set up a Twitter account and followed new people– now what? This is the best time to add some flair to your account. You don't want the users you just followed to come to your profile only to see something that looks abandoned. Simply click "Edit profile" on the right side of your screen below the header on your profile.

First, you'll want to upload two different photos to your profile:

• Twitter profile picture

• Twitter header image

Your profile picture should be something that's clear and distinguishable even as a small icon. It doesn't hurt to upload a larger image, but Twitter will shrink it down to size.

As for the header photo, you will have much more room, so pick something visually striking. All the top businesses that operate on Twitter use appealing photos to engage readers on their profile.

From this same screen, you can update basic profile information. You can catch up on our previous Twitter bio ideas post to get some help. In your profile, you can add a linkable website, your location and a 140-character bio that is linkable and respondent to hashtags and @mentions.

Try to convey what your business or brand is about as succinctly as possible. Don't go overboard with buzzwords or hashtags. You should make your Twitter bio clean and accessible. If you want to update the color scheme of your entire profile, there's also a button for "Theme color." Use a color that matches with your brand so everything seems uniform with your logo and company colors.

4. Start Sending Tweets

Now that everything is ready, you're set to start sending Tweets. Remember, each tweet only allows 140 characters (including Twitter usernames and links). This will help you keep your messages short and sweet, which is much of the allure to Twitter. Aim to be concise without getting too casual.

Your business won't look very professional if you use Internet lingo shorthand.

Abbreviations like TBT (throw back Thursday) and ICYMI (in case you missed it) are perfectly fine for businesses to shorten their message. On the other hand, using words like "ur" for you are and "cuz" for because can seem unprofessional.

Twitter provides two sample posts that you can start with, complete with the #myfirstTweet hashtag. If you want to start with something more unique, you can compose your own first tweet in the text field below the samples. Hit the "Tweet" button and you'll be off and running.

5. Engage

One of the best things about Twitter is how well you can engage with others. You can do this by asking other users for Retweets, followbacks or answers to your questions. You also should continually Tweet from your profile. There's nothing worse than a stagnant Twitter account, and simply put, you won't get much interaction if you give little effort.

Instead, try to schedule Tweets out each day and at specific times that work for your followers. Peak post times always seem to be up in the air, but with Sprout Social's Viral Post feature, you can schedule Tweets to be sent at the time when your followers are most active on Twitter.

It's also good idea to be visual on Twitter whenever possible. This doesn't mean you need an image in every single Tweet, but trends show more social media users engage with visuals than plain text.

Engagement is all about communication. This is why it's important to collaborate with other influencers. With the right connections, you can promote other brands on your profile to get shout-outs on their site. Generally you want to use Twitter to showcase your brand's personality to connect with others.

The more active you are on Twitter, the likelier you'll engage with new users and gain a better following.

6. Integrate With Sprout Social

To completely harmonize your Twitter business experience, connect your profile to Sprout Social to get the most out of your social media campaign. By linking your profile to our platform, you'll be able to:

- Monitor social networks for mentions of your brand

- Create and schedule all of your social posts in advance

- Pull unlimited presentation-ready reports on your success

- Shorten links automatically in the compose box

- Track hashtags associated with your brand

- Automatically post content via RSS from your trusted sites

Through these features, and many more, make sure you have a Twitter management tool you can trust and use to monitor your activity and engagement.

How to create a Twitter marketing strategy for your brand

Twitter marketing has become a complicated channel for even the savviest social marketers.

Twitter has gone from a place for people to share their every thought into a powerful marketing platform that lets brands speak to their audience in real-time. With over 321 million monthly active users, it's easy to see why companies keep using Twitter after all of this time. But it's no longer enough to Tweet about the latest trending topic occasionally.

Like every other social media platform, finding marketing success on Twitter takes strategic planning and intentionality to stand out and keep your audience engaged.

Let's talk about the best ways to use Twitter to market your business, engage with your audience and use the platform to meet your overall social media goals.

How to use Twitter for marketing

At first glance, it may seem like the only way to use Twitter is just to show up and start Tweeting. While you do want to be prepared with plenty of content, there are some strategies that successful businesses use to stand out on the platform and use it to their advantage.

Let's dive into some Twitter marketing tips that any brand can use to make the most out of the platform. If one of these strategies catches your eye, jump right to it with these links.

Audit your Twitter account

If you already have an existing Twitter profile, the first thing to do is to run a Twitter audit.

Take stock of what's working and what isn't working on your Twitter profile by doing an in-depth review of your Twitter analytics. Twitter analytics tools help by allowing you to:

- Analyze hashtag performance

- Analyze individual Tweet performance

- Analyze your individual Twitter audience

Finding out which Tweets are performing the best will give you an idea of the type of content that your audience is most interested in seeing. You can use this information to create a strategy that maximizes the reach and engagement you get on your Tweets and give your audience what they're looking for in your content.

One way to audit your Twitter profile is to look over your analytics manually. Do this by logging into your Twitter profile, navigating to your Twitter Analytics dashboard, selecting the "Tweets" tab and exporting your data for a specific date range.

If you're using Sprout Social to manage your Twitter profile, then you can access your Twitter Analytics right alongside the rest of your social media data on Sprout's Reports tab.

• **Find your Twitter voice**

Audiences on Twitter are looking for brands that Tweet authentically stay true to their voice. It can be easy to jump on the latest trends to appeal to the masses on Twitter but don't do this at the expense of losing your brand voice. While your Twitter presence can be more playful and casual than LinkedIn or Facebook, it should still be authentic and consistent with your brand voice as a whole.

Wendy's is a brand that has nailed its brand voice on Twitter, while still staying true to their brand persona. They're not afraid to have a little bit of fun with current trends while remaining real to who they are as a brand.

An engaging brand voice is essential, but don't jump on trends just for the sake of staying relevant. Twitter users are especially astute to when a brand is being inauthentic in an attempt to generate attention.

Use Twitter hashtags and trends

Tweets with hashtags get almost double the engagement than Tweets without hashtags. This is an exciting statistic, but it doesn't mean that you should load up your Tweets with every hashtag you see. Hashtags are a great way to expose your brand to new audiences who may be interested in what you have to say. Some brands create hashtags for a specific campaign and use that hashtag to label individual Tweets or encourage their audience to share Tweets with that hashtag.

On Black Friday 2014, Kohl's made history by generating tons of engagement among excited shoppers with their #KohlsSweeps Twitter contest. Participants were encouraged to answer trivia questions about recent pop culture events for the opportunity to win gift cards and other prizes. During the weeklong Twitter campaign, Kohl's generated over 450K engagements on their Twitter page, over 370K Twitter mentions and saw over 255MM potential impressions.

The success they saw during that campaign inspired them to create similar campaigns every year to attract attention to their profiles and encourage followers to engage with them on Twitter.

For those times when you can't think of what to post, Sprout's Trends Report helps you monitor hashtags, trending topics and influencers in your industry on Twitter.

Utilize Twitter ads

Using paid ads on Twitter is a great way to reach your audience in a more direct way than waiting for organic reach. Promoted Tweets can expand your reach more quickly.

They also allow people to discover your profile, even if they don't follow your brand or hashtags. When you use a promoted Tweet, your Tweets show up people's timelines who share interests with your audience. You pay a monthly fee as long as you want the promoted Tweet to remain posted. Users can interact with promoted Tweets in the same way they interact with organic content.

The only difference is that promoted Tweets are marked so that users transparently know it's a paid ad.

Progressive uses promoted Tweets to advertise their competitor pricing feature. The ads are true to their brand and are not overly promotional. In fact, they encourage users to go to their website to find the best insurance rates, even if they end up not using Progressive.

Find out when to Tweet

The best overall time to post for the most exposure is Wednesday at 9 a.m. and Friday at 9 a.m. The best days to post are Tuesday and Wednesday. The worst day to post is on Saturday when there is very little engagement on the platform.

For consistent engagement, it's recommended to schedule the majority of your posts for Monday through Friday from 8 a.m.– 4 p.m.

Sprout's ViralPost™ feature looks at your audience's activity and even chooses the most optimal times for you to share updates depending on when your followers are most active.

Even though Tweets remain on your profile forever unless deleted, Twitter moves so fast that something you Tweeted 30 minutes ago may very well be invisible to your followers.

According to a study by Wiselytics, one Tweet has a half-life of 24 minutes. Tweets only reach about 75% of their potential engagement in less than three hours after being posted. This makes posting consistently and at the right time, the key to finding success on Twitter.

Twitter global times to post 2021

The best overall time to post for the most exposure is Wednesday at 9 a.m. and Friday at 9 a.m.

The best days to post are Tuesday and Wednesday.

The worst day to post is on Sunday when there is very little engagement on the platform.

For consistent engagement, it's recommended to schedule the majority of your posts for Monday through Friday from 8 a.m.–4 p.m.

Sprout's ViralPost™ feature looks at your audience's activity and even chooses the most optimal times for you to share updates depending on when your followers are most active.

Schedule Tweets ahead of time

Remember Oreo's iconic Superbowl blackout Tweet?

Power out? No problem. Pic.twitter.com/dnQ7pOgC

— OREO Cookie (@Oreo) February 4, 2013

Tweets like that, which happen at the right place at the right time, can generate massive engagement and virality. But these spur of the moment Tweets are the exception to the rule. The majority of your Tweets should be intentionally scheduled ahead of time for maximum reach.

Knowing the right time to Tweet makes scheduling a consistent flow of Tweets much more manageable. It's important to be as consistent as possible on Twitter. Best practices recommend Tweeting at least once per day. Some brands Tweet up to 15 times per day to stay in front of their audience. Best practices can differ across industries. So it depends on your resources and social media strategy to determine how frequently you can create and publish new content on twitter.

It helps to use a platform like Sprout to visualize when to post your content to make sure posts go out consistently and at the optimal times.

Another way to create engagement with your followers is to ask questions. You can do this by creating a survey on Twitter directly or just asking a question and Retweeting people's responses.

Glossier's Into The Gloss blog regularly poses questions in the form of Twitter surveys to ask their audience questions about their skincare routine. This gives the audience insight into popular opinions about skincare and gives the brand valuable research information that they can use in their content.

Try using a Twitter scheduling tool to stay ahead of your content calendar. Just plain unsure what to post? Here's our guide on what to Tweet.

Engage with your followers

Twitter is all about creating a two-way channel of communication with your audience. It's important to create content that encourages your audience to engage with your Tweets. You also need to make sure that you're engaging with people who are Tweeting about your brand individually. If someone mentions your brand or responds to a Tweet, make sure someone is responsible for responding to their message in a timely fashion. A lot of businesses use Twitter as a way to field support questions. So having a dedicated community manager to handle these requests will help prevent responses from falling through the cracks.

Asana avidly monitors their Twitter mentions and responds to virtually every mention they receive. It doesn't matter if it is a pressing question or general praise. They make sure their users know that they are paying attention to what they are saying.

Sprout's Suggested Replies feature lets you create canned responses for commonly asked questions on social media. These can then be edited further for a personal touch, but they give the people managing the profile behind the scenes a simple way to handle any volume of inquiries they get on Twitter.

Another way to create engagement with your followers is to ask questions. You can do this by creating a survey on Twitter directly or just asking a question and Retweeting people's responses.

Glossier's Into The Gloss blog regularly poses questions in the form of Twitter surveys to ask their audience questions about their skincare routine. This gives the audience insight into popular opinions about skincare and gives the brand valuable research information that they can use in their content.

Set measurable Twitter goals

Your Twitter marketing strategy needs to create measurable goals that will keep your plan on track. Instead of publishing Tweets and hoping for the best, set goals and objectives for Twitter. These goals should help your business meet its overall marketing goals. Goals for Twitter can include:

• Building an engaged following to increase brand awareness

• Generating leads by directing traffic to an offer or email list

• Increasing traffic to your website by posting links to blog content

• Building brand loyalty by providing excellent customer service on Twitter

- Networking with influencers and industry thought leaders to create more connections

Once you've determined your Twitter goals, set aside time every month to measure those goals. Then you can analyze what's working with your strategy and what needs tweaking. It helps to use a social media management platform like Sprout to measure your activity and track your goal's progress.

Incorporate Twitter into your overall social media content strategy

Having a solid social media content strategy lets you visualize what you're looking for from each of your social media profiles and allows you to get the most out of your presence on each network. Large organizations especially may find themselves with a different team managing each social network. Your twitter social media strategy should align with your overall social media content strategy. It's a great idea to use a social media management platform, like Sprout Social, that lets you collaborate across teams and create standards that everyone managing social media can follow.

Every brand's Twitter marketing strategy is going to be unique to their business and industry. Following best practices can build a foundation for a Twitter presence that is memorable and helps your business achieve its goals.

Set goals

Success on any social media platform begins with having clear, measurable goals. There's no way to know if your strategy is having a positive impact on your business unless you understand what you're trying to achieve.

You want to create SMART goals: Specific, Measurable, Attainable, Relevant and Time-bound. So "going viral" doesn't count. These goals should align with your high-level business objectives, and be broken down into measurable indicators of success.

For example, you may want to drive more traffic to your website. Translate that into a SMART goal by aiming to increase your average click-through rate. You can use your baseline click-through rate from your Twitter audit to set a specific achievable goal over a reasonable period of time (say, an increase from 1.5% to 2.5% in three months).

Check out the competition

Isn't doing a great job its own reward? Well, sure. But admit it— you also want to leave your rivals in the dust.

So don't forget to review the Twitter accounts for your industry competitors. Analyzing their social media can help you refine your own, by revealing weaknesses or gaps in their strategy, and ways that you can distinguish yourself.

Assign roles

You need to ensure your accounts are monitored and active, and that someone is replying to direct messages and mentions. Twitter conversations move fast, so it's noticeable to your followers if you're not checking in regularly, and a failure to be responsive and timely will damage your brand.

Busy accounts may need multiple team members monitoring them, like Vancouver's Translink account. Individual team members sign their names, to provide a personal touch to their customer service.

But even if only one person is responsible for your Twitter account, you'll still want to designate a back-up team member so that there are no gaps in coverage.

Ensure everyone is clear on their responsibilities. Too much coverage can provide its own problems, if multiple team members are trying to respond to the same tweets and offering redundant or conflicting answers. A social media management tool like Hootsuite can be helpful for assigning clear roles and responsibilities.

Create guidelines

You need a social media style guide to keep your communications clear and consistent. Guidelines also help you onboard new team members, and prevent mishaps and mistakes on social media.

Your guidelines should be shared with everyone on your social media team, and may include elements of your overall brand style guide, like your tone and details about your audiences.

But it should also be specific to how you use social accounts, including Twitter, with details like:

• Branded hashtags and how to use them

• How and where you use emojis

- How to format links

Every kind of conversation—good, bad, weird— happens on Twitter, so you want to be ready for anything. Criticism is inevitable, especially as your account grows, so you should plan for how to respond to trolls and manage a PR crisis. Remember, it's much better to have those resources and not need them than the other way around.

Make a content calendar

Planning your content takes a little bit of time upfront, but ultimately saves you effort and stress later on. Trust us, you'll be glad you did it when you're not struggling to come up with a witty, original tweet for #NationalDoughnutDay at the last minute.

A social media content calendar is useful for aligning the content you're posting on all your channels, and spotting possible gaps and conflicts that you can address. It also helps you Planning ahead helps seize opportunities for timely or interesting content, like sharing your sustainability practices on Earth Day, or celebrating your female founder on International Women's Day.

When creating your calendar, consider:

• How often you want to post

• The best times to post

• Who should approve posts

A calendar can also help you assess your content and see if you're sharing a balanced mix of tweets. You want to follow the rule of thirds (number 8 on this list): ⅓ of tweets promote your business, ⅓ share personal stories, and ⅓ are informative insights from experts or influencers.

However, you can't set it and forget it. You still need to keep an eye on your Twitter account so you can reply to DMs and mentions and join conversations.

Worried about spending too much time on it? Don't be—you can manage your social media in just 18 minutes per day.

Measure your impact

Once your marketing strategy is underway, you need to be regularly evaluating your efforts and checking your progress against those SMART goals you set.

But the data available to you can be overwhelming—there are a ton of metrics at your fingertips, including vanity metrics that aren't always meaningful. So think about which metrics really matter. Getting a lot of retweets from a funny meme is great—but did any of those translate to conversions or sales?

Collecting meaningful data will help you demonstrate the value of your marketing efforts, and provide insights that will help you refine your strategy over time.

Pros of Using Twitter For Business Marketing

I. Twitter helps you to find your exact target market and those are people who are genuinely interested in what your business is about and its products or services.

II. Twitter allows you to connect and engage with your customers on a really personal level. The interaction is casual and continuous and gives your business a real face and voice to people. This is known as conversational marketing

III. Twitter provides your customers the platform to give you valuable feedback. It's so much easier for them to tweet those 280 characters from the comfort of wherever they are than fill in a feedback form. Your local business can get firsthand reviews in real time enabling you to remain nimble so you can give them what they want.

IV. Twitter provides a sales funnel for your local business, enhancing your ecommerce. You can integrate your twitter account into your business operations by using it to take orders, offer exclusive coupons and promote business and social events.

V. Twitter offers your business a chance to build a community around your brand instead of being a detached entity. This community of followers will become your mini-marketing army through the viral network.

Cons & Disadvantages For Using Twitter For Business

I. It takes some time to set up and gain followers. You need to be organized and patient in the initial stages before you see big results.

II. It can be addictive because you feel chained to your phone all the time, always checking for incoming tweets and then responding to them. It's hard to separate your business life from your personal life and that can affect you socially with family and friends taking second place.

III. It is a double-edged sword and you need to be ready to accept the good and down right nasty tweets about your business.

IV. Bad news travels faster than good news so any mistakes or controversies your business has will travel at lightning speed on Twitter. You need to be extremely quick to do damage control if this happens.

V. Because people always know what you and your business are up to and in some cases your exact location, it could result in security issues if someone maliciously used that information. It is always wise to err on the side of caution and decide what you shouldn't tweet about concerning yourself and your business.

Effective Twitter Marketing tips for your Business success

Here are the 22 Effective Twitter Marketing tips categorized in 5 different sections based on profile, Network, Followers, Engagement, Measuring Results.

Customize your Twitter profile

The very first thing is you should customize your Twitter profile because your profile is the first impression your audience or followers get. Make sure that your profile should be impressive, and it should create interest in your business at a very first glance.

There are some locations you need to customize in your profile.

Choose the right Twitter handle

It is your Twitter user name. It should be easily recognizable and it should be easy to remember as well. Use your company's name as a username of your Twitter account so that people can find your profile easily.

Keep it as short as possible. With the short names, you can reduce the number of characters. It would be easy for someone who mentions you in their tweets or replies to you because of the character limit.

If the brand name is not available, try to get something similar to your brand name.

For example,

• Include your industry – add abbreviations for your industry like "app"

• Use get – add get in your user name. This turns your user name into a call- to –action.

- Use HQ – add HQ in your user name. People assume an account with HQ in user name it the official account for your business

Don't use any punctuation marks in your Twitter handle.

Use the same user name across all your social media profiles that will help people to recognize easily.

Make your profile photo and your header image look great

The header image is the background image of your Twitter profile. It should represent your brand and you can add some unique information which you can't add in your bio. You can add your business slogans in your header image.

Make sure that the image will increase your brand awareness. You can also use your products pictures or pictures related to your brand concept.

Keep your header image simple but make sure that it looks great and it should grab the attention of the people.

The recommended size of a header photo is 1500 px X 500px in JPEG or PNG format. You can also use tools like Canva for creating your header image that offers free templates.

Use your official brand logo as your profile picture that will help to increase brand recognition. The profile picture is always displayed as a circle so make sure that your image's corner won't get cut off when cropped into a circle.

If you like to use company CEOs or initials photos as your profile picture, use some professional headshots.

Write an attractive Bio

Your Twitter bio is right under your profile photo. Usually, it is the first thing when people will look at your profile and it also shows in search results.

It should describe your company within the 160 character limit and it should be engaging & factual. Here accuracy is the key.

Describe exactly what your brand is and tell people what to expect when they follow you.

Write a bio for your target audience and make sure to include brand related keywords in your bio. Add some relevant hashtags in your bio that will help people to find you easily when they search for those terms.

If your brand has multiple twitter accounts, don't forget to include a link for other accounts in your bio so that people can easily find them.

Include your website URL in your Bio

Add your website URL as a clickable link in your Twitter profile is a simple but effective way to get some referral traffic to your website.

Get verified

It is the blue check mark badge that appears next to the account name in profile and also in search results. When people see your account, they will notice that your account was verified by Twitter that will make your brand trustworthy. They trust your content as well. However, Twitter verified account program is currently on hold. They are not accepting any new requests at this time. Please watch out for the update, once it is available you can opt-in for the Verified account program.

Build your Twitter network

Developing your Twitter network is the next step in Twitter marketing strategy. Here are the few tips to build a Twitter network,

• Follow people who are relevant to your business

If you want to people follow you then you have to start following other people first. You have to follow who are relevant to your business. Use Twitter search to find people who are relevant to your brand or business.

You can follow your customers, related businesses, friends, and users who provide useful information.

If you start following people, many of them will following back. So automatically, you can increase your number of followers.

• Find influencers and engage with them

Find relevant influencers, follow them, retweet and comment on their posts, let them know you are listening to them and it is the best way to engage with your influencers.

You can use Twitter search to find relevant influencers and you can also use tools like klear and twitonomy to find influencers.

Find the right influencer is the key here.

Consider the below points when choosing an influencer,

• Number of followers

• Followers growth

- The engagement rate of their posts

- Number of shares on their posts

- Kind of mentions

Send a direct message to the influencers and ask them to promote your business in a tweet. If you have a better connection with them, they would like to respond to you.

They can help you to expand your reach & build credibility and as a result, you can get more followers and you can get more website traffic as well.

In fact, recommendations or suggestions from influencer tweets increase purchase intent by 5.2 times when compared with tweets from brands.

- Use Twitter lists

If you follow a higher number of people, it is really hard to track and interact with everyone's post. You may miss some important tweets and you may miss people those you really want to track.

Twitter lists are the solution for that issue and it is the organized group of Twitter accounts based on a specific category. You can create either a public list or private list that will help to manage your contacts.

You can create separate lists such as,

• Target audience

• Potential customers

• Competitors

• Influencers

• Inspiration people

• Experts

Now you can view the tweets of the list members in a separate Twitter timeline so you can interact with them easily. It is a great way to build a relationship with your target groups.

You can also subscribe to the public lists of other people so that you can view the tweets of the people that you don't follow. This helps to increase your brand reputation.

• Host or participate in Twitter Chats

Host or Participate in Twitter chats is an effective way to expand your Twitter network. It helps you to get more exposure for your profile and it also helps you to make a personal bond with your audience.

Several Twitter users discuss a specific topic at the same time using a shared hashtag is known as Twitter chat.

To host a Twitter chat you need to choose a specific topic or specific question, create a hashtag, and then set a time or date for the chat. While scheduling your chat you need to consider your target audience and the time zones of them.

Now you need to promote your Twitter chat to your followers through Twitter and other social media channels.

Through twitter chats, you can discuss with your audience and ask their opinions as well.

You can also invite special guests and influencers to answer a few questions. This makes your Twitter chat more interesting & more engaging and this will help to reach a new audience as well.

Promote your twitter account outside of Twitter

Advertise your Twitter account is another best way to increase your number of followers. Add twitter button on every page of your website, mention on FaceBook, Ask your email subscribers to follow you on Twitter will help you to promote your Twitter account.

You should give them a reason, describe what benefits they will get for following you on Twitter.

Engage with your followers

Interact with your followers help you to build a personal bond with your followers. So they will like to come back to your profile that increases your brand loyalty. People like your brand when you interact with them.

How to interact with your followers effectively? Here are some tips,

• Follow back people who follow you

Follow back people who will follow you and thank him or her to follow you. You can send a short message or direct message to them that will help to create a personal bond with them.

Check your direct messages regularly and make sure to respond to them.

- Interact with your followers

Review your follower's tweets and if you feel that it is a great content retweet that post or like that post or reply to them.

Remember that it is a platform for conversation, so be sure to connect with your audience always.

When people reply to you respond to them that will help to build your brand image. According to a recent study, 77% of Twitter users feel more positive when the brand replies to their tweet and 60% of customers expect brands to respond to their queries within an hour.

Use your Twitter account for customer service that is the best way to use Twitter for your business. According to Twitter, 85% of SMB Twitter users said that it is important for businesses to provide customer service on Twitter.

When you handling customer service on Twitter, keep your conversation at public whether it is positive or negative. It shows that you provide great customer service and if you need any personal information of your customer then move your conversation from public to privately direct message.

- Use @mentions

Mentions used to draw the attention of other Twitter users. When you mention influencers, followers or other brand's tweet they will receive a notification. It is beneficial to them, they get more exposure. So they likely to return a favor, they may share your post or appreciate the post that helps you to increase your brand reputation.

Make sure that you respond to your customer and media mentions.

- Use retweet with the comment to connect with your followers

Use retweet with comment when you want to share people or brand's interesting or informative tweets instead of just retweet because it adds additional value to your followers.

- Create Twitter polls to interact with your followers

The Twitter poll allows you to question your audience with four options. Twitter polls help you to expand your visibility and it helps to learn about your audience.

All you need to do is craft your questions and select the options. Now click compose tweet box then click the horizontal graph icon. Enter your question and the options in choice boxes.

By default, the poll length is 24 hours, but it also allows you to set the timeframe.

Conduct a poll at peak traffic times so that you can get the highest engagement and use your polls to interact with your followers.

Increase your post engagement

Post engagement plays a pivotal role in the growth of your Twitter account. Here we have listed a few tips to increase the engagement of your Twitter post.

- Share interesting contents with killer headlines

Providing high-quality content is the key to your social media presence. High-quality, compelling contents helps to attract new followers. Share interesting, valuable, informative content with compelling headlines within the character limit 280.

Post relevant trending topics that will make your post more engaging. Make sure that your post should generate curiosity or it should state the benefits.

Create killer headlines that should influence your followers to click the post, retweet the post or like the post.

Share an irresistible quote from your article that helps you to hook the attention of the people. Twitter also allows you to share the links of your blog post or your website. So you can drive traffic to your website through Twitter.

Use pinned tweet to draw the attention of people. Usually, the pinned tweet get more impression so change your pinned tweet often. Make sure to pin the latest blog post of yours that will help you to drive more traffic to your blog post or website.

• Find the peak time for posting

As I mentioned above, the average life span of one tweet is only 18 minutes. So it is necessary to find the optimum time for posting to get maximum engagement otherwise, your post may lose in the crowd.

You have to post when your followers are active on Twitter that helps you to get more clicks and more impressive.

According to a recent study, between 12 pm and 6 pm is the best time for posting on Twitter and your peak time is mainly depends upon your audience. So keep test your posting time and find out which is best for your brand.

You can also use tools like Tweriod to find the best time to tweet. It analyses your followers and shows you when your followers are more active on Twitter. It also shows you the time when your Twitter get the most exposure.

• Use right Hashtags

Hashtags are the best and effective way to expand your reach. If you tweet with hashtags when someone searches for the specific hashtag they can able to see your tweet.

According to Buffer, tweet with hashtags gets 2 times more engagement than those without. Use only minimum hashtags and tweets with 1 or 2 hashtags have 21% higher engagement than those with 3 or more hashtags.

Before adding hashtags in your tweets you need to find out which hashtags are trending. Use Twitter analytics to find your most successful hashtags and use RiteTag tool to find the best hashtags for your tweet.

All you need to do is enter a keyword related to your topic and RiteTag provide a list of relevant hashtags. Now you can choose the most relevant hashtags to your tweet.

You can also check the popularity of the hashtag in the Twitter search bar. When you type the hashtag in the search bar it shows how many tweets use that hashtag in an hour or in a day. It will help you to choose the most popular hashtags so your post will get more visibility.

Use right hashtags is very important here and remember that don't overuse them.

• Use multimedia content to increase retweets

Humans are visual by nature that's why visual contents are very effective to improve the engagement of your post. If you want to grab your audience attention, use multimedia contents like images, infographics, videos, and GIFs in your tweet.

According to Buffer, Tweet with images receives 150% more retweets, 89% more favorites and 18% more clicks. Use relevant and high-quality images only and use a tool like Canva to create your own image.

According to Hubspot, People share infographics 3 times more than any other type of content.

People love videos because it provides all valuable information in a short span of time. According to Twitter, videos are 10 times more engaging and videos are 3 times more likely to be retweeted than GIFs and 6 times more likely to be retweeted than images.

Multimedia contents not only increase retweets but they also drive more traffic to your website.

- Tweet multiple times in a day

You should tweet at least three times a day that will help you to get more exposure and that leads to more engagement. So you can attract more followers as well.

According to Adweek, 92% of companies tweet more than once a day, 42% tweet 1- 5 times a day and 19% tweet 6- 10 times a day.

If you have a global audience you have to Tweet more frequently and make sure to reach the different time zone as possible.

As I mentioned above posting time is depends upon your followers. If it is necessary you can also use Twitter management tools that will make your life easier.

Advertise on Twitter

With the help of Twitter ads, you can target your potential customers and increase your engagement. 3 types of Twitter Ads are there,

- Promoted tweets

- Promoted Accounts

- Promoted trends

Promoted Tweets – it is the best option for advertisers to increase engagement and increase exposure beyond their follower base. Promoted tweets are just like regular tweets, you can retweet, like or reply to that tweet and it is clearly labeled as promoted.

Promoted tweets appear in the timeline of users and in search results.

Promoted Accounts – it is a great option to increase your followers and it appears in suggestions (whom to follow), search results, and user's home timelines. According to Twitter, 85% of Twitter users find promoted accounts useful for discovering new brands.

Promoted account ads target users who are relevant to the advertiser's brand.

Promoted Trends – it appears on the top of the trending topics list on Twitter and it also clearly labeled as promoted. Promoted trends are a good option to promote a specific hashtag.

Promoted trends are visible to all users on Twitter.

Measure your Twitter marketing results

Measuring your Twitter marketing results allows you to estimate your campaign success. With the help of analytics, you can get a clear view of which is work and which is not. So that you can optimize your Twitter marketing in the future for better results.

• Use twitter analytics

Twitter analytics is one of the best tools that you can use to grow your Twitter account. To access your Twitter analytics click your profile image which is in the top right corner then click the analytics.

It is a great place to identify what type of content gets more engagement. Here you can see your top tweets for the month. It will help you to find out popular contents among your audience so that you can continue to create more content around that topic. You can also share this tweet often throughout the month or next month but don't share that tweet exactly. Share it with different images, with quotes or with different words.

Next, you can see your top mentions for the month. Once again thank them that will help you to create a strong personal bond with your audience.

You can also identify who are the most influential people (Top follower) in your audience.

Twitter Analytics also gives you a clear view of the performance of your tweet. It helps you to know the number of impressions, rate of engagement that your tweet has received. You can see that in your tweet next to the like and share button.

USING THE BEST HASHTAGS FOR YOUR SOCIAL STRATEGY.

When it comes to your company's social media marketing, hashtags are pretty much an essential part of your content strategy. These useful little links create the perfect method for finding relevant content to share, giving your audience an easy way to find your content, and grouping together like conversations.

If you've ever been on Twitter or Instagram, you've more than likely seen hashtags in use before. You've likely used hashtags on these platforms (and maybe more!) yourself.

Whether you're a seasoned expert when it comes to hashtags or brand new to social marketing, there's one thing that's certain: you must be able to track your hashtag use. Knowing which hashtags are popular, drive people to find your content and get them talking is essential to your social strategy.

What is a hashtag?

The first hashtag ever was used on August 23, 2007, by a Twitter user named Chris Messina as a means to differentiate between groups of topics within tweets.

Twitter has a long history of adopting user recommendations onto their platform. For example, it was early users who referred to posts on Twitter as "Tweets" and who created @replies and retweets. And the platform just recently added the thread feature as an easy way for people to create stories via tweet.

So it wasn't long until hashtags made their way onto the platform as an official feature. And to say they spread like wildfire would be an understatement.

Hashtags are now commonplace within social media, with platforms like Facebook, Instagram, Pinterest, and LinkedIn eventually implementing the searchable link feature as well.

Now, hundreds of thousands of posts using hashtags are shared on social media daily.

Hashtag best practices

Since hashtags are so common, they can be used in many different ways and on many different platforms. Strategies can range from using your own branded hashtag to encourage your audience to share user-generated content or jumping on trending hashtags to create popular content.

Using hashtags isn't a difficult or advanced part of your strategy. Check out these tips for incorporating hashtags into your social media posts.

From finding the most relevant hashtags to your brand to tracking post performance, Sprout can help you effectively analyze your strategy.

How to find the best hashtags

Finding the best hashtags to use within your social media content is first priority. It's not just a guessing game, although some strategic A/B testing is always a good idea. Each platform has a search bar for you to browse possible hashtags for your content, but those search features don't always give you an idea of how popular the hashtag is.

Using a tool like RiteTag is perfect for gauging interest in hashtags on Twitter and Instagram. You can search for an industry hashtag (for example, #socialmedia) to see how popular it is, and compare it to top related hashtags.

For hashtags on Twitter, RiteTag provides a list of other relevant hashtags, how many tweets use that hashtag per hour, how many retweets that hashtag receives per hour, and how many people are seeing that hashtag per hour. These lists are color coded, with green indicating hashtags that are hot right now, blue for those with lower immediate popularity but a longer lifespan, red for overused hashtags and gray for underused hashtags that are best to avoid.

Although this information is exclusively for Twitter, RiteTag also provides insight into Instagram hashtags based on your search. The information provided isn't nearly as in-depth, but it can still give you a great idea of which hashtags to use within your Instagram post.

You can also check out how many people are using hashtags on Instagram by typing it into the search bar and tapping the Tags section. Click on each tag to see how many posts are using that hashtag, if anyone you follow is using that hashtag, as well as related popular hashtags.

There are also multiple ways to use Sprout Social to find and review the performance of your hashtags. You can use Sprout Listeners to find out how frequently people are talking about your topic, what related terms they're using and what the sentiment around the topic is.

How many hashtags should you use?

The number of hashtags you use depends on which platform you're posting to. A hashtag strategy is not one-size-fits-all when it comes to the various social media networks. In fact, on some platforms, it's better not to use hashtags in your posts.

• Facebook – 1-2 hashtags

Using hashtags on Facebook can help your content gain visibility, attract new community members for your brand and increase your reach overall. The best approach to hashtags on Facebook is to be specific and select 1-2 hashtags that are truly relevant to the content of your post. When Facebook users peruse a hashtag feed rather than their individual newsfeed, they are looking to discover and engage with trending discussions—which is a great time for your brand to be front and center.

• Twitter – 1-2 hashtags

Using up to two hashtags within your tweets can actually double your engagement. We'll talk about different types of hashtags to use (and track) here in a bit, but keep this in mind every time you're about to tweet. However, using more than two hashtags can actually decrease engagement, so maximize your hashtag use at just one or two.

• Instagram – up to 30 hashtags

Instagram posts allow you to share up to 30 hashtags, and in order to maximize reach and engagement, it's actually recommended to use up as many of those 30 as you can. This is why Sprout Social's Instagram Business Profiles Report is so helpful—it shows your outbound hashtag usage during the reporting period and which of the hashtags you've used resulted in the most engagement.

• Pinterest – 0 hashtags

Pinterest made a big announcement in 2018 about how they were bringing hashtag capabilities to its platform, and as of now, there isn't enough data to determine if pins that include hashtags perform better than pins without. However, this platform is similar to Facebook in that its search feature doesn't differentiate when searching for hashtags or keywords. We recommend simply including your keywords within the description, rather than within a hashtag.

• LinkedIn – 2-3 hashtags

LinkedIn has also just recently added hashtags to their platform, but unlike Pinterest, they have done so while also creating a searching method that does focus on hashtags. Which means including up to three hashtags in your company's

LinkedIn post actually can help people to find your content better. Although there's no limit to how many hashtags you can include within your post, we recommend keeping it between two and three.

How to track your hashtags

When you use hashtags, you're increasing reach and engagement on your social media content. Hashtags are one of the most powerful organic social media strategies, but in order to maximize their performance, you need to track your hashtags.

Many brands struggle to identify which social media metrics are most important for their campaigns, and this can also be true of hashtags. With Sprout Social's trends report, it's incredibly easy to see which hashtags your audience is using, how they're performing and what other topics they're mentioned with. This lets you build your content strategy around what your audience is most interested in, rather than just latching on to hashtags that are widely popular but have little relevance to your brand or audience.

Adding those keywords and hashtags into your own tweets can help generate more engagement and buzz around your brand online.

Important hashtag metrics to track

What should you actually look for when knowing a hashtag is working for your social strategy? Here are a few top metrics that can help you determine the success of a hashtag.

- Popularity

How popular is the hashtag that you're using? Hashtags that are used often tend to also be searched for often, so it's a good idea to include hashtags in your post that have proven to be popular. You just want to make sure you're not spamming your followers with irrelevant but trending hashtags just for popularity's sake.

• Reach

How many people actually tend to see the hashtags you're using? If your reach isn't very high, you're probably not using the best hashtags. Try out some new tactics to see if you can increase the eyes on your posts, such as using a tool like Sprout to find related hashtags that are getting a lot of attention.

• Interactions

Not only do you want to make sure people are using and seeing these hashtags, you want to make sure people are also interacting with them. Posting hashtagged content that gets users Retweeting and sharing will expand the reach of your campaign.

- Users

Who, specifically, is using the hashtags? You want to make sure that you're seeing users that are within your target audience using and searching for the same hashtags that you are so your message resonates.

Which hashtags to track

There are several different types of hashtags that you'll be using in your social media content. Here's know how and why to track each one.

- Content hashtags

Content hashtags are essentially the keyword hashtags that you tag at the end or inside of your post. Going back to the example we used in this article when searching for relevant hashtags, #socialmedia would be considered a content hashtag. Your company's industry hashtags would also be considered content hashtags.

You'll want to track these so that you know which are the most popular content hashtags to use when sharing content like a blog post or other industry-related news. Check tools like RiteTag to understand which content hashtags are the most popular and are seen the most often.

- Branded hashtags

A branded hashtag is a hashtag that your company has created and promotes as a way of tagging your company directly. For example, @KITKAT uses #KITKAT as a way to promote their product across Twitter.

Many times, a branded hashtag is simply your company name or your company name + a keyword like your product or service.

Tracking this hashtag is a great way to see how many people are talking about your business specifically. You'll want to use a monitoring tool like Sprout Social's Discovery feature to keep tabs on conversations surrounding your branded hashtag.

• Trending hashtags

You can easily find out which hashtags are trending by checking out your sidebar on Twitter's desktop website or by tapping the magnifying glass icon on the smartphone app. The platform itself will let you know how many people are talking about each of the trending hashtags so you'll know ahead of time if it's worth incorporating into your content.

Using trending hashtags in a way that's relevant to your content and appropriate to your brand voice can help increase viral attention on your post. Be sure to check your Twitter report to see how the impact differs from using your regular hashtags.

- Event hashtags

If your company is hosting an event, if your team members are attending an industry event or conference, or if you're tweeting about a large event going on nationally or internationally (i.e., big sporting events), be sure to incorporate the event's hashtag in all of your posts about that event.

Live tweeting is a great way to generate traction around your event and event hashtag and increase reach and engagement on your content.

Monitoring your Sprout Social Twitter report is a great way to determine how much engagement your event chatter generated around your business and if you were able to increase your reach, follower count, and more. Doing some social listening is a great way to check in on how many people were also using the same event hashtag or discussing related topics.

- Campaign/ad hashtags

A campaign/ad hashtag is a hashtag that your company creates for a specific campaign, launch, or digital advertisement. This helps to generate buzz around one specific thing that your business is doing.

For example, Starbucks will use different hashtags each time they're promoting a new drink.

If your company is trying to promote a new product or service launch or is running a specific campaign, creating a new hashtag for your company and your customers to use is a great way to get the buzz out. These hashtags are also great for tracking since they're so specific to each campaign's focus compared to your recurring brand or content hashtags.

Social listening is a great strategy for monitoring use of your campaign/ad hashtags and how well they're catching on with your audience.

Twitter analytics tools to amplify your strategy

Twitter is an excellent platform to engage with friends and follow your favorite industry experts. But if you want to use it to promote your business or your personal brand, you have to do more than just broadcast tweets every once in a while. It's not enough to blindly share your new blog posts and product updates, either.

A great Twitter strategy requires in-depth Twitter analytics to understand what's working and what isn't. Are your posts generating any clicks? Is there any content format that's performing better than others? Which topics do your followers love the most? At what time of day are your followers most active?

Analyzing all these aspects of your Twitter performance is essential to measure your social ROI, so you know you're not just wasting your time on tweets that don't deliver. It also helps you identify what kind of improvements to make and where you're falling behind. That's why it's worth investing in the best Twitter analytics tools to keep track of how your posts are performing.

In this post, we've put together six of the best tools to analyze your Twitter efforts. Some of these tools will also help you identify trends and keep track of how your competitors are performing. All of this gives you a better idea of how to improve your Twitter strategy for sustainable growth.

1. Sprout Social

We top this list for good reason—Sprout is an industry-leader in Twitter publishing, engagement, and, you guessed it, analytics tools.

From reach and engagement to link clicks and hashtags performance, Sprout makes it simple to identify and optimize your top-performing tweets. You can also gain insight into the Tweets that resonate most with your audiences with Premium Analytics. With both templated and customizable reporting options, it's easy to analyze and share your most important Twitter data. You can prepare custom monthly, quarterly or yearly reports based on your business' reporting structure to measure the success of your social strategy.

Get a clear overview of your overall impressions, or use the Sent Messages Report to dig deeper into the individual performance of each tweet. Tailor these reports further with Premium Analytics. By adding and removing the right metrics you can extract actionable insights from your Twitter data.

If you want more than a one-dimensional look at your performance, generate a Twitter Comparison Report to see how you stack up against the competition.

Use the Twitter Trends Report to discover trending topics and hashtags across the platform. Shape your content strategy based on these trends to increase visibility and engagement. This report even helps you narrow down on influencers and brand advocates who are talking about you.

Combined with other publishing, engagement and strategy management features, Sprout's Twitter analytics will give you an all-in-one solution to manage all your Twitter activities in one place.

Best for: Anyone who needs an all-in-one solution to manage their social media.

2. Union Metrics

Union Metrics helps you visualize your social data through colorful graphics that are easy to understand even for novice marketers. It's an analytics-only service that offers a comprehensive social analytics solution, including Twitter account analysis.

You get to monitor your Twitter activity in real time and receive the latest performance reports to stay up to date. Its Twitter analytics offerings also include keyword listening, campaign reporting and competitor analysis. These features let you audit your overall performance on the platform and optimize it if necessary.

Besides its paid analytics tool, the company also offers a few free Twitter tools including the Twitter Snapshot Report. This gives you an overview of your brand performance on Twitter—perfect if you ever need a quick report for audits and presentations.

There's also the Twitter Assistant tool, which provides you with recommendations customized according to your account for free. You'll be able to discover the best time to tweet, the best types of content to post and which hashtags drive the most impressions.

This tool is an excellent addition to your social media marketing toolset if you're using a free publishing service that doesn't offer much in terms of analytics.

Best for: Businesses that need a basic analytics solution at a budget price to use alongside their publishing software.

3. Tweepsmap

The better you know your audience, the better you'll get at engaging them. Tweepsmap is a great Twitter analysis option to gain a deeper understanding of your Twitter community and how to better engage them.

This tool lets you map your followers and identify key demographics including gender, language and profession. Use these insights to segment your community into relevant categories. This gives you leverage in creating content and campaigns that will appeal to different segments of your audience.

Understand your Twitter activity in terms of how many followers you're gaining (or losing). Identify your most influential followers if you ever need to launch an influencer marketing or brand advocacy campaign.

While the free plan is solely focused on analytics, Tweepsmap also offers basic publishing features like the Tweet Scheduler for Premium users.

Best for: Companies and individuals who need a more comprehensive community analytics solution to go with their basic social media analytics software

4. Keyhole

Another analytics-only service, Keyhole lets you track conversations and understand audience sentiments around them. Get a better idea of how people on Twitter feel about your brand or your competitors by tracking keywords and hashtags.

You can use this Twitter analytics app to discover actionable insights that will help you optimize your performance. It helps you learn how to craft more engaging tweets and when you should share them to get the most engagement.

You won't even have to manually track your top-performing content and what time you posted them. This tool automatically recommends an Optimal Post Time for your account using visuals that are easy to process and understand.

Keyhole also simplifies the process of deciphering your competitors' Twitter strategies. It helps you keep track of their account growth rate and activity, as well as their engagement data. You'll then be able to use these insights to identify which tactics are most effective in your niche and which ones you should avoid.

Best for: Businesses that want to conduct competitive analysis to optimize their Twitter strategies

5. Native Twitter Analytics

Most social media marketers would already be familiar with the native analytics dashboard on Twitter. It's one of the best free Twitter analytics tools available, making it an excellent option for those who aren't yet ready to invest in a premium solution.

It gives you all the basic analytics data to understand how your tweets are performing. This includes an overview of how many impressions you garnered within a certain timeframe. So you'll be able to keep track of when there were any spikes or drops in performance and what could have led to those changes.

This dashboard also lets you keep track of other performance metrics like number of mentions and profile visits from this dashboard.

Make the most of the audience demographics data to understand your community on a deeper level. Here, you will see the breakdown of your audience based on several factors such as interests, household income, occupation, buying behavior and interests.

These insights will come in handy when you're brainstorming content ideas that would appeal to your audience. You'll also be able to use them for developing effective Twitter marketing campaigns that target specific demographics.

Best for: Businesses that need data but aren't ready to invest in premium Twitter analytics tools

What comes next?

There are tons of Twitter analytics tools to choose from—whether you're willing to invest in a premium service or you need a free solution. But with too many options, it can be confusing to begin your search. Hopefully, this list helped you identify the best tools to focus on based on your needs.

CONCLUSION

Twitter is an effective marketing tool to reach your target audience, increase brand awareness, drive more traffic, and increase sales. It is a very powerful tool for your customer service as well. So twitter plays an unavoidable role in any business's marketing strategy.

With the right tactics, you can increase your visibility and engagement. I hope all the above ways and tips will help you to achieve your goal.